Clive Five

A Numberline Lane book
by
Fiona and Nick Reynolds

Clive Five was very excited. He had just bought a brand new ball. It was bright red.

He went out to his garden and practised some of his football tricks.

He tried keeping the ball in the air, bouncing it on his head.

"One, two, three, four, five," he counted, before it fell to the ground.

Then, Clive Five tried bouncing the ball on his knee.

"One, two, three, four, five," he counted again, before it dropped.

Clive Five thought it would be much more fun if he had someone else to play with.

Just then, Kevin Seven came walking past.

"Hello, Kevin Seven!" said Clive Five.

"Hello, Clive Five!" said Kevin Seven.

"Would you like to come and play with my new red ball?" asked Clive Five.

"Yes please," said Kevin Seven.

They started to play in the garden, kicking the ball to each other.

"Wouldn't it be good if we could organise a football match for all the other Numbers?" said Clive Five.

"Great idea! I will go and tell everyone to meet at the Add Pad," said Kevin Seven.

Soon, everyone from Numberline Lane had gathered on the grass.

They were very excited about playing in a football match.

Gus Plus agreed to be the referee.

The Numbers had to sort themselves into two teams.

Clive Five and Kevin Seven said that they would be one team and everyone else could be in the other team.

"No, no, no!" said Gus Plus, "That wouldn't be fair. There are ten numbers. If there are two in one team then there would be eight in the other team, because two add eight is ten."

"Maybe if Walter One, Nora Four and I make one team," suggested Suzie Two, "would that be fair?"

"No, no, no!" said Gus Plus, "That's still not fair."

"If there are three in one team then there would be seven in the other team, because three add seven is ten."

"To be fair, we need the same number on each team."

"Hmmm!" thought Clive Five, "This is difficult to work out."

Gus Plus helped by putting all the Numbers into pairs that made a total of ten. Walter One stood with Nigel Nine. Suzie Two stood with Katy Eight. Hebe Three stood with Kevin Seven. Nora Four stood with Nick Six.

"Who can I stand with to make ten?" asked Clive Five.

"I think we need another Clive Five," said Nora Four.

"That's right," said Gus Plus, "because five add five makes ten."

"So we need five on each team," said Clive Five, "A five-a-side football match!"

The game got off to an exciting start.

Walter One scored the first goal, kicking the ball straight through Jenny Ten.

Then Suzie Two got the ball and kicked it to Nora Four, who headed the ball to Nigel Nine.

Poor Nora Four was feeling quite exhausted with all the running around.

She asked if she could stop playing and watch for a while.

All this time, Linus Minus had been sitting on a bench watching the match with his football rattle.

He really wanted to join in.

Clive Five had said that he wasn't allowed to play.

He was worried that Linus Minus might take the ball away.

Gus Plus said that Linus Minus should be allowed to play if he promised not to take anything away.

Linus Minus was so excited that he could hardly control himself.

He ran as fast as he could towards the ball, and with one enormous kick he scored a goal straight through Katy Eight!

"Hooray," cheered all the Numbers as Gus Plus blew the final whistle.

One goal each – what a fair way to end the match!